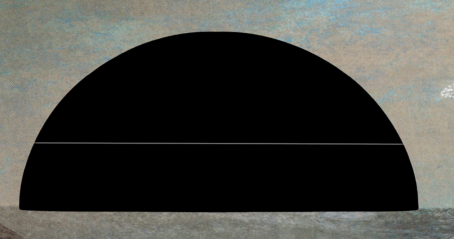

Text by Patricia Hegarty
Text copyright © 2024 by Little Tiger Press Limited
Cover art and interior illustrations copyright © 2024 by Britta Teckentrup

All rights reserved. Published in the United States by Doubleday, an imprint of Random House Children's Books, a division of Penguin Random House LLC, 1745 Broadway, New York, NY 10019. Originally published in the United Kingdom by Little Tiger Press, London, in 2024.

DOUBLEDAY YR with colophon is a registered trademark of Penguin Random House LLC.

Visit us on the Web! rhcbooks.com

Educators and librarians, for a variety of teaching tools, visit us at RHTeachersLibrarians.com

Library of Congress Cataloging-in-Publication Data
Names: Teckentrup, Britta, illustrator. | Hegarty, Patricia, author.
Title: Family : a peek-through picture book / illustrated by Britta Teckentrup ; text by Patricia Hegarty
Description: First American edition. | New York : Doubleday Books for Young Readers, 2025 | "Originally published in the United Kingdom by Little Tiger Press, London, in 2024." | Audience: Ages 3–7 | Summary: "Peek-through holes on each page allow readers to discover how animal families live in the wild." —Provided by publisher.
Identifiers: LCCN 2024005999 | ISBN 978-0-593-90200-4 (trade)
Subjects: LCSH: Familial behavior in animals—Juvenile literature.
Classification: LCC QL761.5 .T43 2024 | DDC 590—dc23

MANUFACTURED IN CHINA   10 9 8 7 6 5 4 3 2 1
First American Edition

Random House Children's Books supports the First Amendment and celebrates the right to read.

LTK/1800/1276/0924

FAMILY

A Peek-Through Picture Book

Illustrated by
Britta Teckentrup

Doubleday Books for Young Readers

On the savanna, whatever the weather,
The animals must stick together.

A herd of elephants cross the trail,
Steadily walking, trunk to tail.

When it's time to leave the nest,
Tiny chicks will face a test.

Learning quickly, they must try
To spread their wings and start to fly.

Deep in the ocean, when danger lurks,
Dolphins have a trick that works.

They keep their calves safe and sound
By forming a circle all around.

Love can be shown in many ways,
A warning look, a watchful gaze.

A gorilla sits among the trees,
Grooming her young for ticks and fleas.

In a land of snow and ice,
An emperor penguin does not think twice.

He resolutely stands his ground,
His egg at his feet, safe and sound.

**Baby lions stay close to their mother.
This pride will care for one another.**

**Mom keeps all her family near,
Safe from threats and free from fear.**

Way down deep, near the ocean bed,
Peaceful creatures are nurtured and fed.

A floating pack of manatees
Are kept from the dangers of the seas.

**Floating happily near the sands,
Otters hold each other's hands.**

As they sleep, they gently sway,
Knowing they won't drift away.

High up among the rustling trees,
A soft sound catches on the breeze.

An owl is hooting his courtship song,
Seeking a mate for his whole life long.

**A wolf hunts everywhere for food
To feed his playful, hungry brood.**

Under the light of a silvery moon,
His precious cubs will be eating soon.

Where ice and snow lie thick and deep,
Polar bear cubs fall asleep.

And here they'll stay all through the night,
Together, as one, until first light.

As the sun comes up on a brand-new day,
The elephant herd is on its way.

They trek beneath the rising sun.
Their journey together has just begun.